On the Line

The Story of the Greenwich Meridian

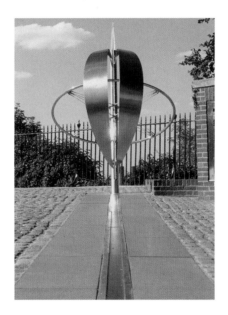

Graham Dolan

ROYAL
OBSERVATORY
GREENWICH

NATIONAL
MARITIME
MUSEUM

Where East Meets West

The Royal Observatory, Greenwich is famous around the world as home to both the Prime Meridian and to Greenwich Mean Time. Just as the Equator separates the Northern and Southern Hemispheres, the Prime Meridian separates the Eastern and Western Hemispheres. But why does the Meridian pass through Greenwich and not Brussels, New York or the Pyramids? Why does it pass through an observatory and not a centre of government? Did the Meridian exist before the Royal Observatory? And what is the link between the Meridian and time? The answers lie in the history of navigation and the part played in this by the Astronomers Royal and others at Greenwich.

▼ Looking northwards along the Meridian in the Observatory courtyard.

▲ Many people come to the Observatory to stand astride the Meridian and have a photograph taken with one foot in the East and the other in the West. (In this south-facing picture, East is to the left and West is to the right.) The Meridian runs from the North Pole to the South Pole passing through the Airy Transit Circle, a telescope located in the building behind.

◀ George Airy, the seventh Astronomer Royal (1835–1881), after whom the Airy Transit Circle is named. His predecessors were: John Flamsteed (1675–1719), Edmond Halley (1720–1742), James Bradley (1742–1762), Nathaniel Bliss (1762–1764), Nevil Maskelyne (1765–1811) and John Pond (1811–1835).

▼ The Meridian is marked at night by a laser beam. The beam originates from a point above the Airy Transit Circle. Depending on atmospheric conditions, it can be seen from more than 20 km away. The floodlit building is the oldest part of the Observatory. It was built in 1675 by Sir Christopher Wren on the instruction of King Charles II.

Lines of Latitude and Longitude

Lines of latitude and longitude form the grid system used on globes, maps and charts. Latitude is a measure of how far North or South somewhere is from the Equator; longitude is a measure of how far East or West it is from the Prime Meridian. Lines or parallels of latitude all run parallel to the Equator. Lines or meridians of longitude all converge at the Earth's North and South Poles. The north–south line passing through any particular point on the Earth's surface is known as the 'local meridian'.

Latitude and longitude are both measured in degrees. Each degree of latitude corresponds to a distance on the Earth's surface of about 111 km. Each degree of longitude, however, corresponds to a distance that varies with latitude. The distance is about 111 km at the Equator, reducing to 0 km at the poles. Although the Equator is an obvious zero point from which to measure latitude, there is no equivalent from which to measure longitude. In 1884, an international conference decided that the meridian line on which the Airy Transit Circle stood at Greenwich should be adopted as the Prime or zero Meridian for the world. Other agreements were made about the way in which the world would record time and, as a result, the system of time zones came into being.

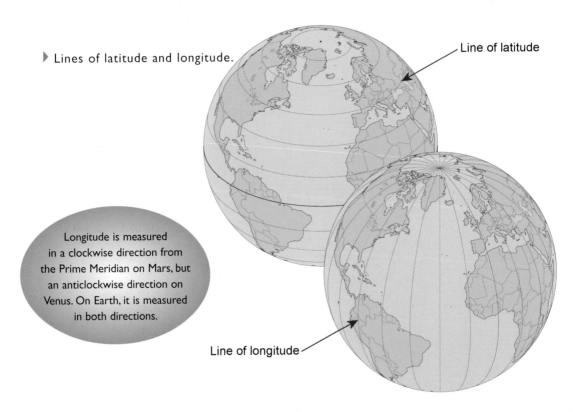

▶ Lines of latitude and longitude.

Line of latitude

Longitude is measured in a clockwise direction from the Prime Meridian on Mars, but an anticlockwise direction on Venus. On Earth, it is measured in both directions.

Line of longitude

4

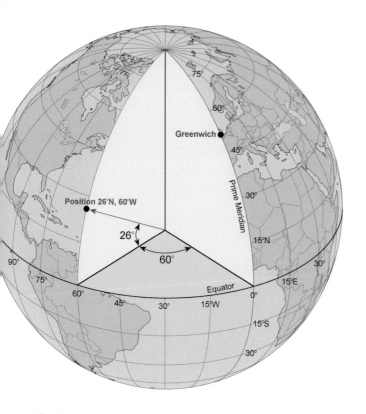

◀ Latitude and longitude are measured in degrees from the centre of the Earth.

▼ Declination and right ascension give the position of a star in the same way that latitude and longitude give the position of a place on the Earth. Stars move across the sky in a similar way to the Sun. They reach their highest point when crossing the local meridian.

Astronomers are able to determine a star's declination and right ascension by measuring the time at which this happens, along with its altitude at the same moment. The use of a telescope aligned along the meridian allowed these measurements to be made with greater accuracy. Once a star's position is known, it can be used to determine the exact local time on each occasion it re-crosses the meridian.

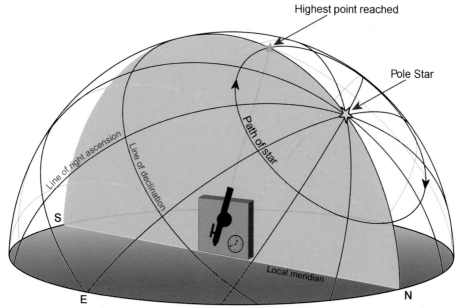

Time Zones

The world's countries fit loosely into a grid of 24 time zones – each 15° of longitude wide. The system is centred on the Prime Meridian and based on Greenwich Mean Time. The time in each zone is an hour ahead of the zone to its west and an hour behind the zone to its east. When the time is 2.00 p.m. in Italy, it is 1.00 p.m. in the UK but 3.00 p.m. in Greece.

Most of the large countries that straddle several times zones, such as the USA, Australia and Russia, keep quite strictly to this system; others such as China use one standard time across the whole country. Regional

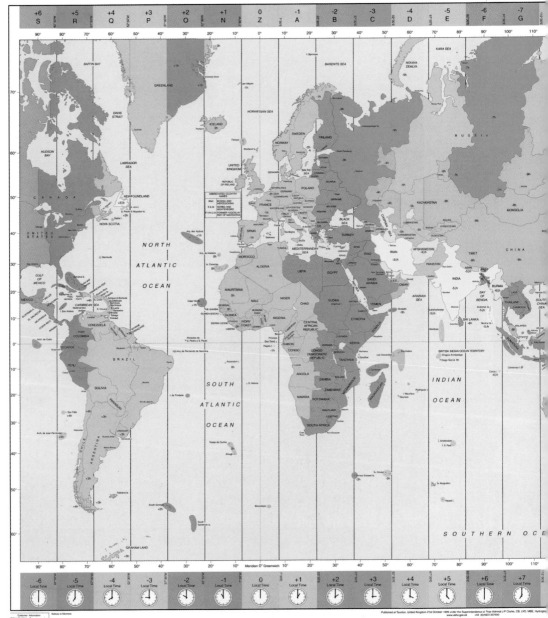

variations are usually decided by political or geographical factors. Occasionally, as in the case of India, time differences are based on half hours rather than whole hours.

Some countries make use of Summer Time (also known as daylight saving time). Clocks are put forward by an hour in the spring and back again in the autumn. In the European Union, clocks are put forward on the last Sunday in March and back again on the last Sunday in October.

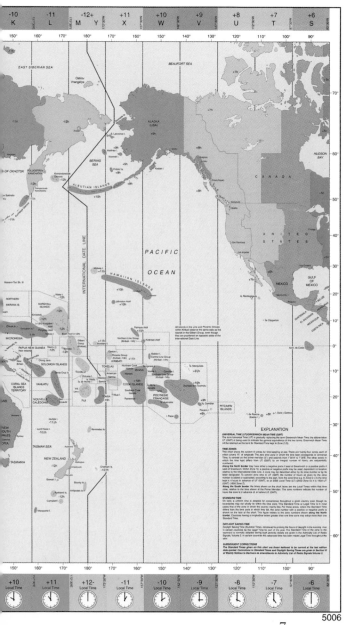

One group of islands in the Republic of Kiribati in the mid Pacific was separated from the rest by the International Date Line until 1995 when its government decided to move the line eastwards.

◀ The world's time zones. The International Date Line lies on the opposite side of the world to the Prime Meridian at longitude 180°. People crossing in a westerly direction skip forward a day. Those crossing in an easterly direction repeat the day.

Longitude From Time

When Christopher Columbus sailed across the Atlantic in 1492 there was no reliable way of measuring a ship's longitude once out of sight of land. Much of the world remained unexplored and charts were inaccurate and incomplete. In later years as the trade routes opened up, these problems became more serious. Journeys often took longer than expected and could end in complete disaster if a ship got lost or ran aground. Some maritime nations offered large rewards to anyone who could find a reliable way of measuring longitude at sea. The two solutions, when they came, were linked to the measurement of time and both were successfully developed in England.

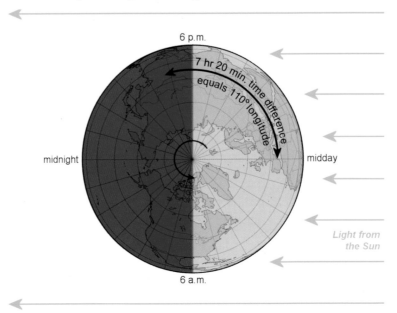

6 p.m.

7 hr 20 min. time difference equals 110° longitude

midnight

midday

Light from the Sun

6 a.m.

▲ When it is midday on one side of the Earth, it is midnight on the other. Each 15° of longitude is equivalent to a time difference of one hour. So to find out how far east or west he was from home, all a sailor had to do was compare the local time (which could be found from observations of the Sun or the stars) with the time back home at exactly the same moment. How, though, was a sailor in the 1500s to know what the time was at home? One suggestion was to take a clock to sea. However, in the 1500s and the 1600s clocks were too inaccurate. The first pendulum clocks came into use in 1657. Although they worked well on land, the movement of a ship prevented them from working properly at sea. As a result, people looked for other solutions.

An observer's latitude is roughly equal to the altitude of the Pole Star in the Northern Hemisphere and can be obtained by observing the Southern Cross in the Southern Hemisphere.

▶ George Anson was so uncertain of his longitude when rounding Cape Horn in the *Centurion* in 1741 that he ended up wasting a lot of time zigzagging back and forth in search of a safe passage. During this time, 70 sailors died of scurvy.

▼ Charts in the 1600s were considerably less accurate than those available today. This one was produced in 1666.

Greenwich and the Moon

As long ago as 1514, it had been proposed that the Moon could be used as a clock for measuring longitude at sea. Some 161 years later, in 1675, the Royal Observatory was built at Greenwich in order to turn what was still just a hypothesis into a working solution – a task so complex, that it took another 91 years to complete.

The Moon's position against the background of stars changes in a complicated but predictable manner. In 1766, the fifth Astronomer Royal, Nevil Maskelyne, published the first *Nautical Almanac*. It contained a set of tables showing where an observer would see the Moon during 1767 if he could be positioned at the centre of the Earth. The Moon's angular position relative to nearby bright stars was listed at three-hourly intervals of Greenwich time.

To determine his longitude, a sailor had to measure the angle between the centre of the Moon and a listed star – the lunar distance – along with both their altitudes. This was done using a marine sextant. Next, he had to calculate what the Moon's position would have been if it had been viewed from the centre of the Earth. The time this corresponded to at Greenwich was then determined from the *Nautical Almanac*. From the difference between this and his local time, he was able to calculate his difference in longitude from Greenwich.

While the Lunar Distance Method was still under development, the British Government, in 1714, set up the Board of Longitude. It offered a prize of up to £20,000 to anyone who could find a means of measuring longitude at sea to the nearest half degree. It was won by the British clockmaker, John Harrison, who, at his fourth attempt, managed to build a timekeeper that would keep good time at sea. Now called H4, it was first tested on board ship in 1761. Today, time-keepers designed for navigation at sea are called marine chronometers.

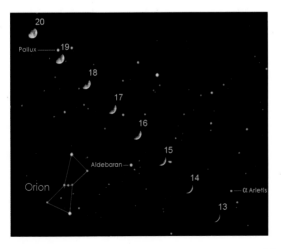

◀ The Moon moves against the background of stars by about 13° each day. Its position and phase as viewed from Greenwich at 7.00 p.m. on successive days in March 2005 has been marked. The stars Aldebaran, Pollux and αArietis are all listed in the *Nautical Almanac*.

The *Nautical Almanac* was developed from lunar tables published by the German astronomer, Tobias Mayer. This page is from the first edition, which sold for 2 shillings and 6 pence.

A marine sextant.

MARCH 1767. [35]					
Days.	Stars Names.	Noon.	3 Hours.	6 Hours.	9 Hours.
		° ′ ″	° ′ ″	° ′ ″	° ′ ″
3		40. 56. 54	42. 37. 12	44. 17. 4	45. 56. 30
4		54. 7. 14	55. 44. 3	57. 20. 26	58. 56. 22
5		66. 49. 20	68. 22. 37	69. 55. 29	71. 27. 56
6	The Sun.	79. 3. 57	80. 33. 57	82. 3. 35	83. 32. 50
7		90. 53. 41	92. 20. 51	93. 47. 43	95. 14. 15
8		102. 22. 23	103. 47. 25	105. 12. 3	106. 36. 26
9		113. 34. 50	114. 57. 58	116. 20. 56	117. 43. 44
7	α Arietis.	43. 16. 5	44. 50. 10	46. 24. 0	47. 57. 33
8		25. 9. 1	26. 33. 51	27. 59. 4	29. 24. 39
9	Aldeba-	36. 36. 33	38. 3. 27	39. 30. 25	40. 57. 28
10	ran.	48. 12. 58	49. 40. 6	51. 7. 13	52. 34. 21
11		59. 49. 55	61. 17. 2	62. 44. 9	64. 11. 16
12	Pollux.	28. 52. 41	30. 20. 0	31. 47. 25	33. 14. 56
13		40. 33. 53	42. 1. 53	43. 29. 58	44. 58. 8
14		15. 18. 50	16. 47. 18	18. 15. 54	19. 44. 37
15		27. 10. 9	28. 39. 36	30. 9. 8	31. 38. 47
16	Regulus.	39. 8. 42	40. 39. 3	42. 9. 30	43. 40. 5
17		51. 14. 48	52. 46. 8	54. 17. 36	55. 49. 11
18		63. 29. 12	65. 1. 42	66. 34. 23	68. 7. 13
19		21. 58. 7	23. 31. 17	25. 4. 44	26. 38. 29
20	Spica ♍	34. 31. 3	36. 6. 18	37. 41. 51	39. 17. 37
21		47. 20. 34	48. 57. 59	50. 35. 40	52. 13. 40
22		60. 27. 51	62. 7. 37	63. 47. 42	65. 28. 6
23		28. 3. 45	29. 46. 14	31. 29. 2	33. 12. 11
24	Antares.	41. 53. 7	43. 38. 21	45. 23. 56	47. 9. 50
25		56. 4. 18	57. 52. 12	59. 40. 25	61. 28. 57
26	β Capri-	16. 37. 26	18. 22. 45	20. 8. 5	21. 55. 58
27	corni.	31. 0. 41	32. 50. 58	34. 41. 35	36. 32. 30
28	α Aquilæ.	52. 37. 35	54. 5. 44	55. 34. 56	57. 5. 6

Distances of)'s Center from ⊙, and from Stars west of her.

F 2

Harrison's prize-winning timekeeper, H4.

The Royal Observatory was built on the foundations of an old castle, using mainly recycled materials from Tilbury Fort and the Tower of London. It cost just over £520, and was paid for by selling off old gunpowder that had passed its use-by date.

One Line for the World

The 1770s marked a turning point in navigation. After years of uncertainty about their longitude at sea, sailors now had two ways of measuring it. Both worked by measuring time differences. The Lunar Distance Method, using a sextant and the *Nautical Almanac*, always gave time differences from the Greenwich Meridian. Marine chronometers, on the other hand, allowed the time difference from any chosen town or city to be measured directly.

For many years, different countries measured longitude from different meridians. The French and Algerians, for example, used the Paris Meridian; the Swedes measured from one that passed through Stockholm. By the 1880s, many people could see the advantages of measuring from a single meridian. As a result, the International Meridian Conference took place in 1884 in Washington, DC. The Greenwich Meridian was chosen to become the Prime Meridian of the world. There were several reasons for this; the main one being that nearly two-thirds of the world's ships were already using charts based on it.

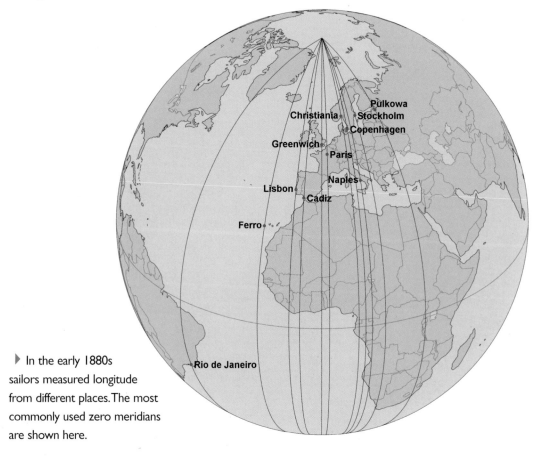

▶ In the early 1880s sailors measured longitude from different places. The most commonly used zero meridians are shown here.

◀ The Prime Meridian is defined for the world by the Airy Transit Circle at the Royal Observatory.

▼ A detail from a map published by Rumbold Mercator in 1595. The longitude scale along the bottom shows the location of Greenwich to be about 21° east of his zero meridian. It also shows England to be about 9¼° of longitude wide – nearly 25 per cent wider than it actually is.

The length of the metre was defined in 1791 as one ten millionth of the distance from the North Pole to the Equator on the Paris Meridian. Since 1983, it has been defined as the distance travelled by light in a vacuum during a time interval of 1/299,792,458 of a second.

Countries on the Line

From Pole to Pole, the Prime Meridian covers a distance of 20,000 km. In the Northern Hemisphere, it passes through the UK, France and Spain in Europe and Algeria, Mali, Burkina Faso, Togo and Ghana in Africa. The only landmass crossed by the Meridian in the Southern Hemisphere is Antarctica.

▲ Although the Prime Meridian passes through nine countries in three continents, for nearly two thirds of its length it passes over the sea.

No matter in which direction people standing at the South Pole look, they will always be facing north.

▶ The position of the Meridian is marked on the N149 at Chalandray about 25 km west of Poitiers in France. Further south, it passes through the vineyards of Bordeaux.

14

▲ 'Walking on water' above the North Pole at the northern end of the Meridian.

▲ Unicycling 'around the world' at the South Pole.

▲ The Meridian passes through the western end of the main breakwater in the port of Tema in Ghana
– the most southerly point of land on the Meridian in the Northern Hemisphere.

On the Line – From Greenwich, South to Peacehaven

From coast to coast, the Meridian in the UK covers a distance of about 330 km. Its position is marked by a variety of monuments and signs. Although the Meridian passes through the grounds of two hospitals, three cemeteries and more than 10 schools, it crosses just one motorway, the M25, which it cuts in two places.

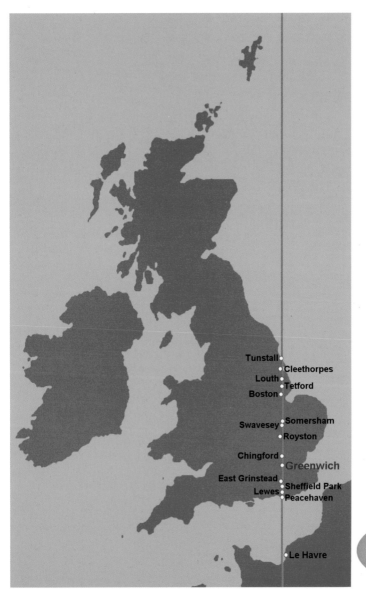

◀ The Prime Meridian passes through the centre of just a few towns in the UK. Louth is the most northerly one directly on the Line, while Peacehaven on the coast is the most southerly.

▲ The inhabitants of Peacehaven erected this obelisk in 1936 to commemorate the reign of King George V and to mark the town's position on the Prime Meridian.

Two trains collided on the Meridian at Hither Green in South London. The accident happened on 12 March 2001.

The Meridian passes along the terrace of East Court Mansion in East Grinstead, whose grounds are open to the public. The Line is marked in several places. This picture was taken looking northwards along the Line towards a stone meridian marker that was erected to celebrate the start of the year 2000.

The 'Meridian Stone' on Lane End Common dates from 1953. It is about 1 1/2 km south of the meridian marker at Sheffield Park Station.

Steam trains on the Bluebell Line cross the Meridian as they enter and leave Sheffield Park Station. A sign marking its position is located just to the right of the signal.

In Lewes, the Meridian is marked on the pavement of the A277, just a few metres to the east of The Meridian pub.

On the Line – From Greenwich, North to Tunstall

Travelling north from the Observatory, the Meridian crosses the River Thames, passes up the Lee Valley and then runs close to the Roman road, Ermine Street, before entering the sparsely populated fenlands of Cambridgeshire and Lincolnshire. It continues across the Lincolnshire Wolds before crossing the Humber Estuary into Yorkshire.

▲ The Meridian passes through Meridian School and Greenwich power station before clipping the site of the Millennium Dome. The siting of the power station was controversial because smoke from the chimneys affected observations made along the Meridian.

◀ The transmitter south of Royston on Periwinkle Hill stands almost astride the Meridian.

The Meridian crosses several golf courses. At some holes, golfers tee off in one hemisphere and sink their putts in the other.

◀ The village of Swavesey is halfway along the British section of the Meridian. The Line passes through the grounds of its primary school and is marked on the southern edge of the village playing field.

▲ The village clock stands just a few metres to the west of a pair of Meridian markers in Somersham High Street.

▼ The Meridian crosses the graveyard of Tetford Church. Some of the bodies buried there have their head in the Eastern Hemisphere and their feet in the Western Hemisphere.

▲ There are two Meridian markers in Eastgate in the market town of Louth. This one is the oldest and dates from 1948.

▼ Driving over the Meridian on the B1362 in Yorkshire.

▲ Jogging across the Meridian near Cleethorpes on the Humber Estuary. The signpost gives the distance to the two Poles and a number of other locations.

▶ The Meridian leaves the coast near the village of Tunstall, Yorkshire. A pillar marked its position from February 1999 until January 2003 when erosion of the soft clay cliffs caused it to topple onto the beach. The remains of the pillar can be seen here.

The Prime Meridian at Greenwich

The Royal Observatory at Greenwich was active until the early 1950s when its astronomers moved away from the light and smoke pollution of London to a new site at Herstmonceux in Sussex. It was only when Flamsteed House (the oldest of the Observatory buildings) opened to the public as a museum in 1960 that the Prime Meridian was marked out on the ground in the courtyard.

▲ This photograph from around 1910 shows how the wooden gates prevented the public from peering into the courtyard of the working Observatory. At that time there was no public access and no reason to go to the expense of marking the position of the Prime Meridian on the ground. Its position was defined for the world by the Airy Transit Circle.

During the Millennium celebrations, some visitors took their photographs in front of a copper lightning conductor running vertically down the wall close to the Observatory entrance. A sign alongside explaining how the new millennium would start for the world on the Prime Meridian lead them to believe that the lightning conductor was the Meridian itself!

▶ A brass strip marked the Meridian from 1960, when this picture was taken, until 1992. It was kept polished with the help of boy scouts and girl guides. A 1960 press release stated: 'The Meridian is marked across the court-yard so that visitors can indulge in the conceit of being photographed standing with a foot in each hemisphere.'

▼ The newly restored Meridian Building was opened to the public in 1967. The courtyard paving also dates from this period.

▼ Although not marked in the Observatory courtyard until it became a museum, the Meridian had been marked on the path outside much earlier. These visitors to Greenwich Park in the 1920s are about to cross the Line from East to West. Many visitors still cross the Meridian at this point today.

Other Meridians at Greenwich

In order to minimise disruption to the observing programme at Greenwich, whenever a new meridian telescope was obtained it was generally set up along a line a little to the east or west of those currently in use. This resulted in a number of different Greenwich meridians. Until the twentieth century the fact that they were in slightly different places was unimportant in navigational and timekeeping terms, as the time difference between them was too small to measure with the clocks and instruments that were then available.

Today, the Meridian defined by the Airy Transit Circle has been superseded for most practical purposes by the International Reference Meridian. Although based on the Prime Meridian, its position is not defined by a telescope. It is determined instead from measurements to orbiting satellites and from distant radio sources such as quasars. It runs about 102 m to the east of the Airy Transit Circle.

◀ The tip of the arrow indicates where the Observatory's first transit telescope was located. Set up by the second Astronomer Royal Edmond Halley, it defined the Greenwich Meridian from 1721 until 1750.

Due to their convergence at the Poles, the Airy and Bradley Meridians are about 30 cm closer where they leave the Yorkshire coast than they are at Greenwich.

▲ Standing astride the Bradley Meridian. This meridian was established by the third Astronomer Royal, James Bradley, in 1750. It was the principal meridian in use when the Ordnance Survey started in 1791 and is still the meridian on which Ordnance Survey maps are based. It was also the meridian in use at the time of the first *Nautical Almanac*. The telescope that defines its position is located underneath the now sealed roof openings. It is about 6 m to the west of the Prime Meridian and 37 m to the east of Halley's.

◀ The alignment of the telescopes along their meridian was regularly checked by observing stars such as Capella that circle the Pole Star and cross the meridian twice each day. This granite obelisk was erected in 1824 by the sixth Astronomer Royal, John Pond, on the Bradley Meridian at Chingford for a similar purpose. It was positioned by direct observation from Greenwich about 18 km to its south.

▶ The Airy Transit Circle defined the Greenwich Meridian from 1851 and the Prime Meridian of the world from 1884. It was used for determining Greenwich Mean Time until 1927, about the time this picture was taken. The Prime Meridian is sometimes referred to as the Airy Meridian.

The Global Positioning System

The Global Positioning System (GPS) was developed by the USA as a military navigation system. It consists of 24 solar-powered satellites, orbiting at an altitude of about 20,200 km, and a network of ground stations. The satellites transmit low power radio signals, which are picked up by a receiver. The receiver locates four or more of the satellites, calculates the distance to each, and uses the information to calculate its own location. To do this, the receiver also has to know where the satellites actually are. This information is stored in its memory. The US Department of Defence monitors the exact position of each satellite and transmits updates as part of each satellite's signal.

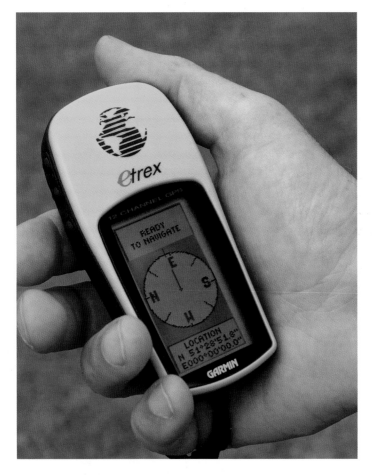

◀ Basic civilian GPS receivers are a similar size and price to a mobile phone.

▼ The 24 GPS satellites are arranged in six orbital planes with four satellites in each plane. At least four satellites are always 'visible' in the sky from anywhere on Earth.

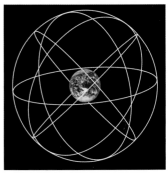

The Earth's tectonic plates are moving relative to one another at about the rate at which fingernails grow.

▼ GPS receivers calculate their distance from each satellite by timing how long the signals take to arrive. Each satellite contains several atomic clocks and each receiver an ordinary quartz clock. The quartz clocks are synchronised from information transmitted by the satellites. A GPS timing error of just 1/100 second would result in an error on the ground of about 3000 km! The Lunar Distance Method allowed a sailor to measure his position to around the nearest 30 km. GPS is over three thousand times more accurate, with an average error of just a few metres.

▼ The Earth wobbles very slightly on its axis. Over the past 100 years, its North Pole has spiralled about 15 m towards New York. In the same period, the tectonic plates that make up the Earth's crust have moved several metres in different directions. The relationship between the latitude and longitude grid of 1884 and places on the ground is continuously changing. However, in most cases the relative distances between places on the same tectonic plate have not changed. GPS allows users to select a co-ordinate system appropriate to their region of the world and to existing national mappings. The measured latitude and longitude of a place can change by more than 200 m if a different co-ordinate system is selected. The map shows the Earth's tectonic plates in the position they were in 250 million years ago.

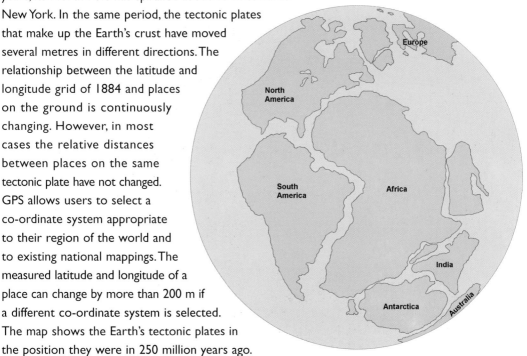

The Time on the Line

Telescopes aligned along a meridian were used at Greenwich for time finding and for measuring star positions until it ceased to be a working observatory in the 1950s. Similar instruments were used in other observatories around the world. Time is no longer determined at Greenwich, but astronomers still measure the Earth's rotation and play an important part in determining civil time.

▶ The Earth spins round on its axis once each day. As it does so, the telescope sweeps across the sky. The telescope was used like the 'hour' hand of a clock, while the stars were used like the 'numbers and marks' around the dial. Observations were taken during the day as well as at night. The 67 stars used as 'clock stars' at Greenwich during the year 1850 are marked, along with some of their names and the times they crossed the meridian.

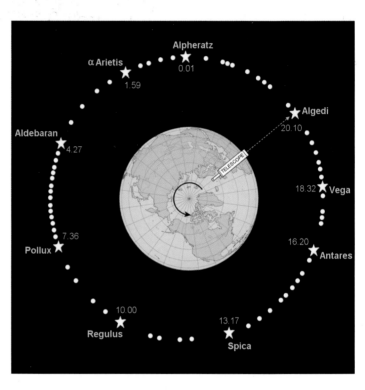

◀ From 1927 until its closure as a working observatory, a small telescope located in the building on the left was used for determining the time. The picture shows the building a few years after it was constructed in 1891. It was demolished in the 1950s.

▲ Not all clocks are marked in hours. This one is marked in degrees – 360° being the equivalent of 24 hours.

▲ This telescope on the Bradley Meridian defined the principal Greenwich Meridian and Greenwich Mean Time from 1816 until 1850 when it was replaced by the Airy Transit Circle. In a 'normal' day, the Earth turns through about 361°; the extra degree compensating for the fact that the Earth is also orbiting the Sun. The time for the Earth to spin round once on its axis through 360° is called a sidereal day and is about 4 minutes shorter than a normal day. The clock was regulated to show sidereal time. The telescope was used to determine its error. Greenwich Mean Time was calculated from the sidereal time and transferred to other clocks at the Observatory and around the country. The theoretical difference between Greenwich Mean Time determined by this telescope and Greenwich Mean Time determined by the Airy Transit Circle to its east would have been about 1/50 second and too small to measure by any clock when the Airy Transit Circle came into use in 1851.

The Earth is gradually slowing down. Days today are about two hours longer than they were 400 million years ago.

Greenwich Mean Time

The further west you are, the later the Sun rises and the later it sets. When a sundial in Greenwich is showing 9.00 a.m., one to its west in Cardiff will show 8.47 a.m. The time indicated by a sundial is called the 'local apparent time'.

The time when the Sun crosses over the local meridian is called the 'local noon'. The length of each 'natural' day measured from one local noon to the next varies slightly through the year. The longest is about 51 seconds longer than the shortest. This difference is due to the tilt of the Earth on its axis and its elliptical orbit around the Sun, rather than variations in the rate at which it is spinning. Each day measured by a clock has the same length and is equal to the average or mean length of a natural day. This is where the word 'mean' in 'mean time' comes from. When natural days are shorter than average, a clock will seem to lose time compared to a sundial. When they are longer, it will appear to gain.

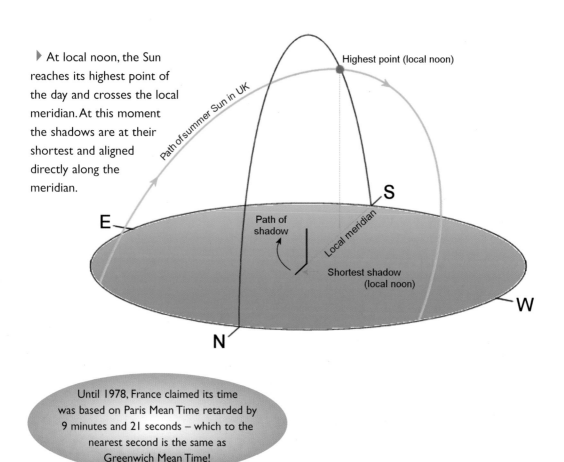

▶ At local noon, the Sun reaches its highest point of the day and crosses the local meridian. At this moment the shadows are at their shortest and aligned directly along the meridian.

Highest point (local noon)

Path of summer Sun in UK

S

Path of shadow

Local meridian

Shortest shadow (local noon)

E

W

N

Until 1978, France claimed its time was based on Paris Mean Time retarded by 9 minutes and 21 seconds – which to the nearest second is the same as Greenwich Mean Time!

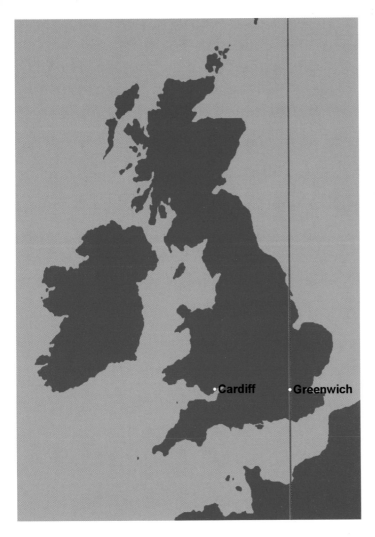

Cardiff •Greenwich

◀ Until about 150 years ago people set clocks and watches to show local mean time, Greenwich Mean Time being 13 minutes ahead of Cardiff Mean Time. Nowadays, everybody within a country or time zone sets their clocks and watches to the same time for civil purposes.

▲ Portable sundials have a compass so that they can be aligned along the local meridian before being read.

▶ Sundials are always aligned along the local meridian. The abbreviations a.m. and p.m. stand for ante meridiem and post meridiem, meaning before and after the Sun has crossed the meridian. Today they are used to mean before or after 12.00 noon, which is not quite the same thing, as time is no longer measured with sundials. In mainland England, the Sun can cross the local meridian as early as 11.37 a.m. or as late as 1.29 p.m., depending on the location and time of year.

Modern Timekeeping

Although the Royal Observatory is the home of Greenwich Mean Time, the public time service at Greenwich came to an end at the start of the Second World War, when it was moved to Abinger in Surrey and later to Herstmonceux in Sussex. In the twentieth century, clocks were developed whose rates were steadier and more stable than that of the spinning Earth, which is gradually slowing down. In 1967, without changing its length, scientists changed the definition of a second from one based on 1/86,400 of an average day to one based on the properties of caesium atoms. In 1972 a new time scale, Co-ordinated Universal Time (UTC) replaced Greenwich Mean Time as the basis of international timekeeping. For most practical purposes they can be regarded as the same thing, because UTC is adjusted so that it always remains within 0.9 seconds of Greenwich Mean Time.

Time today comes from the averaged readings of about 200 atomic clocks located in laboratories around the world. The International Earth Rotation Service in Paris co-ordinates information from satellite tracking stations and other sources about variations in the Earth's spin rate. Adjustments of a second at a time are made to keep the UTC time scale in step with the gradually slowing Earth. In the first 20 years of UTC, a total of 22 days had an extra second added to them. These extra seconds are called leap seconds.

◀ The laser satellite tracking station at Herstmonceux is one of a number of satellite tracking stations located around the world. By using a laser beam to measure the distance to orbiting satellites in known and stable orbits, variations in the Earth's spin rate can be measured, along with tidal movements of the land and continental drift.

◀ This vertically mounted telescope was used for time measurement at Herstmonceux from 1957 until the 1980s. The only stars visible were those that crossed the local meridian directly overhead. To allow for the longitude difference from Greenwich, a correction of minus 1 minute 21.0785 seconds had to be made to obtain Greenwich Mean Time.

Clocks are currently being developed whose theoretical accuracy is equivalent to losing or gaining no more than one second in the lifetime of the universe.

▶ This atomic clock, dating from the 1990s, was designed to keep time to the equivalent of 1 second in 1,600,000 years.

First published in 2003 by the
National Maritime Museum, Greenwich,
London, SE10 9NF.

ISBN 0 948065 50 8

3 4 5 6 7 8 9

A CIP catalogue record for this book is available from the British Library.

Commissioned by Rachel Giles.
Photographic work by Leah Desborough, Darren Leigh, Josh Akin
and Lisa Macleod of the NMM Photo Studio.
Artwork by Anton Vamplew, Andrew Sinclair and Harry Ford.
Editorial, design and production by The Book Group.
Cover design by Mousemat Design Ltd.

All images © National Maritime Museum apart from:
Pages 6–7 © Crown Copyright 2003. Reproduced from Admiralty chart 5006
by permission of the Controller of Her Majesty's Stationery Office
and the UK Hydrographic Office (www.ukho.gov.uk).
Page 14 (bottom) photography by Pieter van der Merwe.
Page 15 (top) © Ann Hawthorne/B&C Alexander,
(middle) B&C Alexander, (bottom) © Geoslides.
Pages 20,23,30 © NMM by permission of PPARC and of the Syndics of
Cambridge University Library, Royal Greenwich Observatory Archives.
Cover images: Andrew Sinclair and Leah Desborough, NMM.
Frontispiece: NMM

Printed and bound in China by 1010 Printing International Ltd.